ANIMAL

GALÁPAGOS PENGUIN OR EMPEROR PENGUIN

BY ERIC GERON

Children's Press
An imprint of Scholastic Inc.

A special thank you to the team at the Cincinnati Zoo & Botanical Garden for their expert consultation.

Library of Congress Cataloging-in-Publication Data
Names: Geron, Eric, author.
Title: Hot and cold animals. Galápagos penguin or Emperor penguin / Eric Geron.
Other titles: Galápagos penguin or Emperor penguin
Description: First edition. | New York : Children's Press, an imprint of Scholastic Inc., 2022. | Series: Hot and cold animals | Includes index. | Audience: Ages 5–7. | Audience: Grades K–1. | Summary: "NEW series. Nonfiction, full-color photos and short blocks of text to entertain and explain and how some animals with the same name can survive in very different environments"—Provided by publisher.
Identifiers: LCCN 2021044798 (print) | LCCN 2021044799 (ebook) | ISBN 9781338799521 (library binding) | ISBN 9781338799538 (paperback) | ISBN 9781338799545 (ebk)
Subjects: LCSH: Galapagos penguin—Juvenile literature. | Emperor penguin—Juvenile literature. | Penguins—Juvenile literature. | Habitat (Ecology)—Juvenile literature. | BISAC: JUVENILE NONFICTION / Animals / Penguins | JUVENILE NONFICTION / Animals / General
Classification: LCC QL696.S473 G47 2022 (print) | LCC QL696.S473 (ebook) | DDC 598.47—dc23
LC record available at https://lccn.loc.gov/2021044798
LC ebook record available at https://lccn.loc.gov/2021044799

10 9 8 7 6 5 4 3 2 1 22 23 24 25 26

Printed in the U.S.A. 113
First edition, 2022

Book design by Kay Petronio

Photos ©: cover left and throughout, cover right and throughout: Tui De Roy/Minden Pictures; 4: Mike Wilkes/NPL/Minden Pictures; 5: Samuel Blanc/Biosphoto; 8 center: Sue Flood/NPL/Minden Pictures; 10 bottom left: Tui De Roy/Minden Pictures; 10 right: mrtekmekci/Getty Images; 12–13: Michael S. Nolan/BluePlanetArchive; 14–15: Mark Spencer/Auscape/Minden Pictures; 16 center: mrtekmekci/Getty Images; 16 bottom: Jim McMahon/Mapman ©; 17 right: Jim McMahon/Mapman ©; 18–19: Michael S. Nolan/BluePlanetArchive; 24–25: Tui De Roy/Minden Pictures; 26–27: Stefan Christmann/Nature Picture Library; 28 main: Michael S. Nolan/BluePlanetArchive. All other photos © Shutterstock.

CONTENTS

MEET THE PENGUINS

Galápagos penguins and emperor penguins are very different. Galápagos penguins live in the warm Galápagos Islands. They are able to find ways to keep cool in the hot weather.

GALÁPAGOS PENGUIN

FACT

Galápagos penguins hold their flippers out to stay cool on land.

Emperor penguins live mostly in chilly Antarctica. They are able to find ways to stay warm in the freezing-cold weather.

FACT

Emperor penguins huddle together to stay warm against the cold.

PANTING BEAK

The Galápagos penguin opens its beak to pant like a dog as a way to cool off.

WHITE TUMMY

The white feathers on its tummy make the Galápagos penguin blend in to keep safe from predators swimming below.

GALÁPAGOS PENGUIN CLOSE-UP

A Galápagos penguin can weigh 4 to 6 pounds (2 to 3 kg).

It has black and white feathers, with a black mark on its chest and a white line from its eye to its chin.

BLACK FEATHERS

A Galápagos penguin's black feathers help it **camouflage** in the water from predators flying above.

FLIPPER POWER

These strong flippers make Galápagos penguins superfast swimmers.

FACT

Galápagos penguins are the only penguins found north of the **equator**.

EMPEROR PENGUIN CLOSE-UP

An emperor penguin can weigh 55 to 90 pounds (25 to 41 kg).

It has black and white feathers, with orange and yellow on its head, neck, and chest.

MOUTH SPIKES

There are spikes, inside their mouths and on their tongues to hold food in place before they swallow it whole!

Emperor penguins can grow up to 3-½ feet (1 m) tall!

FIRM FLIPPERS

Emperor penguins have flippers instead of wings in order to help them swim.

FINE FEATHERS

Tightly packed feathers keep them warm and waterproof in their freezing-cold climate.

SMOOTH BODY

An emperor penguin's body is perfectly designed to zip quickly through the water!

NICE NAILS

Sharp nails on their feet grip the ice to keep emperor penguins from slipping.

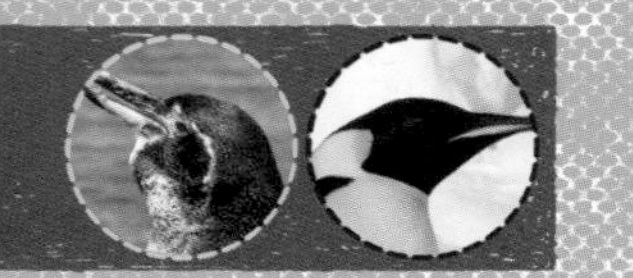

TINY AND TALL

GALÁPAGOS PENGUIN

Galápagos penguins are the second-smallest penguin. Emperor penguins are the largest penguin.

Galápagos penguins are tiny compared to tall emperor penguins—about half their height!

FACT

Even though penguins are birds, they cannot fly.

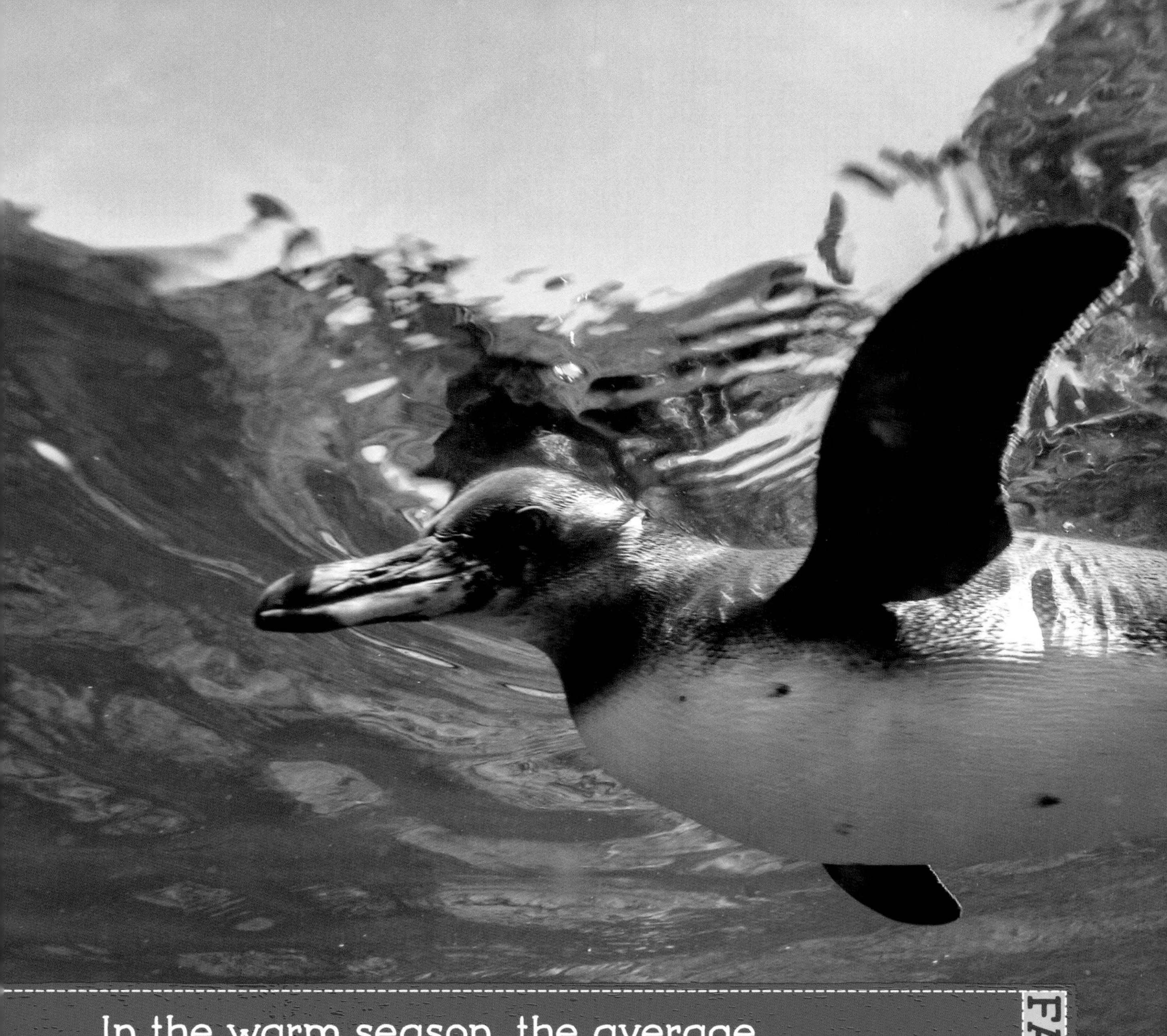

FACT

In the warm season, the average temperature in the Galápagos is 78°F (26°C).

ISLAND PARADISE

Galápagos penguins live in the Galápagos Islands. The islands are in the Pacific Ocean off of the coast of South America. The weather is warm and **tropical** there. Galápagos penguins beat the heat by staying cool in the ocean water.

PENGUINS ON ICE

Emperor penguins live on the icebergs and ice cliffs of Antarctica. They mostly waddle or slide along the ice on their stomachs. Although emperor penguins can move on land, they spend almost all their time in the frosty water. Emperor penguins live in **colonies**, or groups.

FACT

In the winter, the average temperature in Antarctica is −40°F (−40°C)!

BIRD TALK

Galápagos penguins and emperor penguins live in very different places. Galápagos penguins stay in the same location forever.

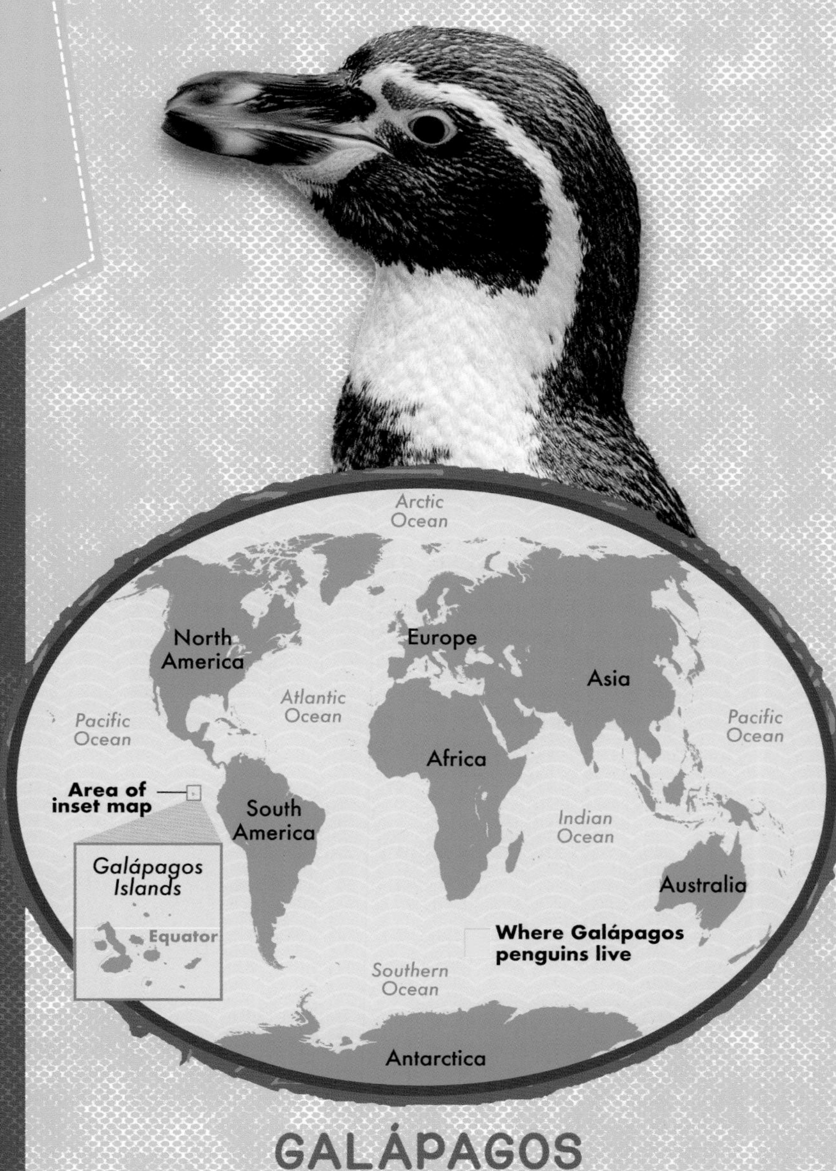

GALÁPAGOS PENGUINS

However, each March, the emperor penguins travel great distances to reach their nesting grounds.

Both Galápagos penguins and emperor penguins use their voices to communicate. They use unique calls that sound like squawking to recognize each other.

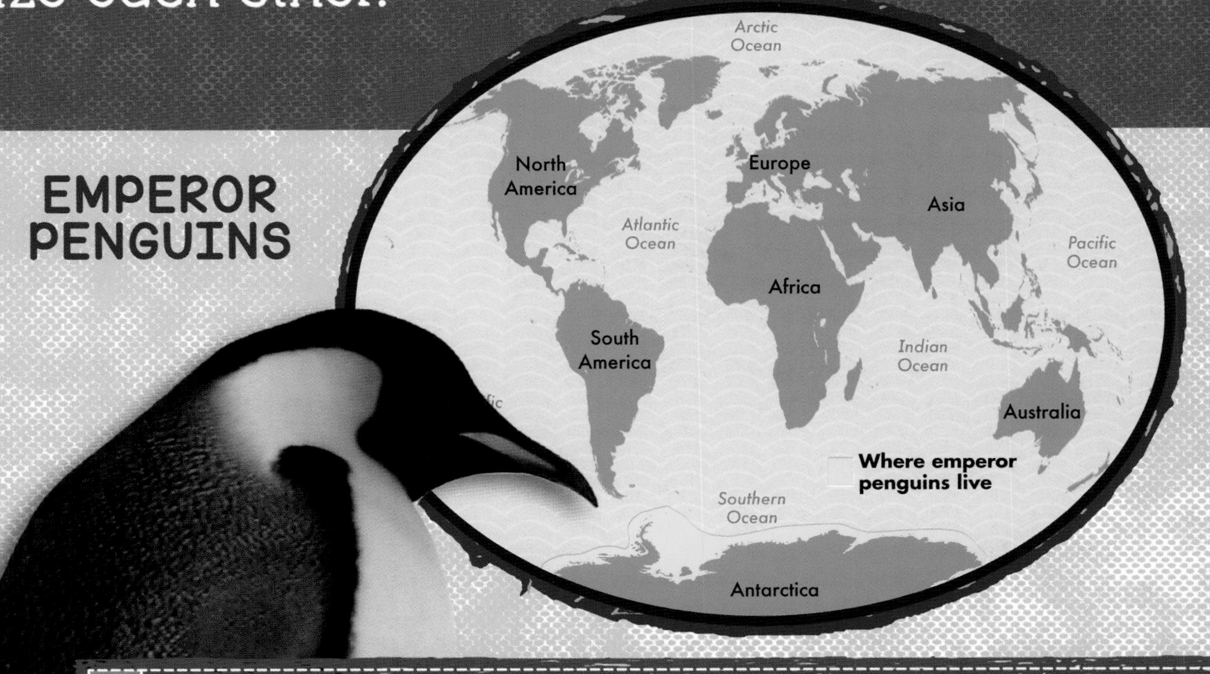

FACT Penguins' body movements to communicate are called "displays."

On average, Galápagos penguins can only hold their breath for less than three minutes!

FISH FOOD

Galápagos penguins are carnivores. A **carnivore** is an animal that eats meat. Galápagos penguins mostly eat small fish like sardines and anchovies. Sometimes they will eat **crustaceans** like shrimp. Galápagos penguins mostly hunt in groups and find food close to shore. They can only dive short distances.

GOING FISHING

Emperor penguins are also carnivores. Emperor penguins mostly eat fish, **krill**, and squid that live around the edges of ice shelves. An emperor penguin is the world's deepest-diving bird! They can reach depths of 1,850 feet (564 m). Emperor penguins swim through the water to find their food.

FACT

Emperor penguins can stay underwater for up to 20 minutes at a time!

A PENGUIN PARTY

Galápagos penguins eat all year long. The amount of food that Galápagos penguins can eat depends on the ocean. Sometimes the ocean carries a lot of food, but sometimes there's none. This lack of food is one of the reasons why Galápagos penguins are **endangered**.

FACT

Sharks, sea lions, hawks, and fishing nets are the main threats to the Galápagos penguin.

Emperor penguins must eat a lot of food in order to make it through the long winter. They gain body weight to help keep warm during the coldest months of the year. During this time, they do not need to eat. A male emperor penguin can survive for more than 100 days without food!

FACT

Killer whales, leopard seals, and giant seabirds are some of the emperor penguin's predators.

Galápagos **chicks** grow special brown and white feathers to protect from sunburn.

FACT

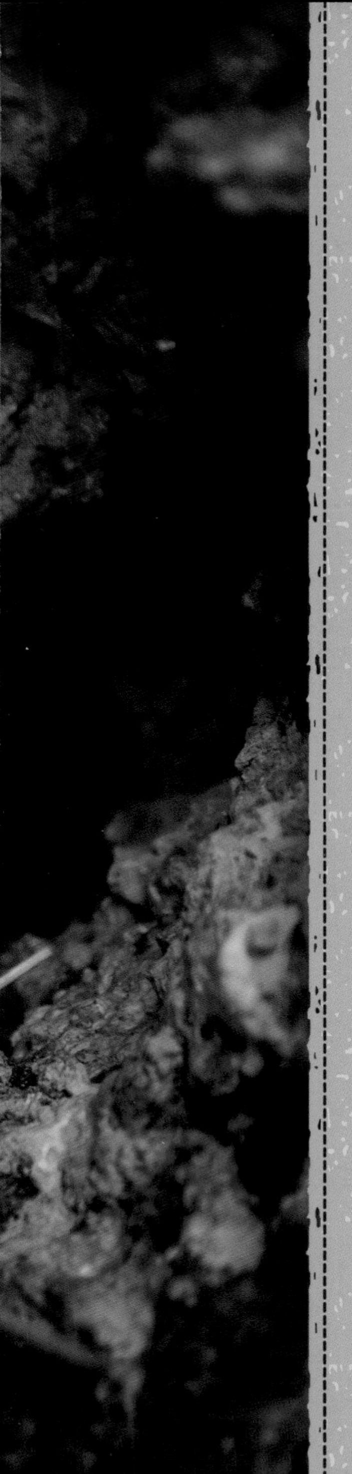

COOLING EGGS

Galápagos penguins live in small colonies. They lay eggs all year long and try to keep their eggs nice and cool. Galápagos penguins build their nests in shaded places like tunnels and caves. The female usually lays two eggs, and both parents take turns caring for the eggs. Baby penguins are called chicks. When the chicks are 3 to 6 months old, they leave the nest and find their own food.

WARMING EGGS

The female emperor penguin usually lays one egg in May or June, which is wintertime in Antarctica. She gives the egg to her male partner to keep safe and warm. The male balances the egg on his feet and under his tummy until it hatches. The chick will leave the colony and head to sea to fish once it is around 4 months old.

FACT A group of emperor penguin chicks is called a **crèche.**

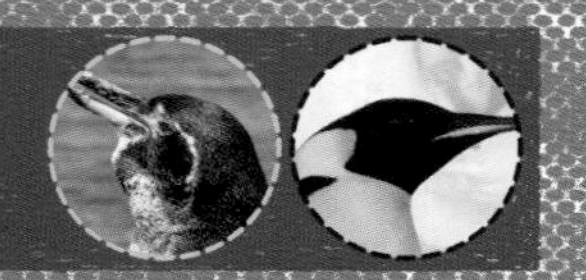

SWEETEST CHICKS

GALÁPAGOS PENGUIN CHICK

Although they grow up to be different adult penguins, Galápagos penguin chicks and emperor penguin chicks are not so different. They are both covered in fluffy feathers.

Both penguin chicks **molt**, or shed their fluffy feathers, for new adult feathers. Galápagos penguin chicks and emperor penguin chicks are both taken care of by their mother and father.. Each chick makes a sound like a whistle to ask for food.

FACT

Penguins have short legs, so it's easier for them to **waddle** instead of walk.

YOU DECIDE!

If you had to choose, would you rather be a Galápagos penguin or an emperor penguin? If you like warm tropical weather and don't mind the hot sun, maybe you would choose to be a Galápagos penguin. If you like snowy weather and sliding around on the ice, you may prefer being an emperor penguin!

There are 18 different types of penguins.

GLOSSARY

camouflage (KAM-uh-flahzh) – to disguise something so that it blends in with its surroundings

carnivore (KAHR-nuh-vor) – an animal that eats meat

chick (chik) – a baby penguin

climate (KLYE-mit) – the type of weather in an area

colonies (KAH-luh-nees) – a group of animals that live together

crèche (kresh) – a group of emperor penguin chicks

crustaceans (kruh-STAY-shuhns) – spineless animals with hard bodies such as crabs, lobsters, and shrimp

endangered (en-DAYN-jurd) – to be at risk of no longer existing

equator (i-KWAY-tur) – an imaginary line around the middle of the Earth that is an equal distance from the North and South Poles

krill (kril) – a small shrimplike ocean creature

migrate (MYE-grate) – to move from one area to another, usually during different seasons

molt (mohlt) – to shed (old feathers or fur)

predator (PRED-uh-tur) – an animal that lives by hunting other animals for food

tropical (TRAH-pi-kuhl) – of or having to do with the hot, rainy area of the tropics

waddle (WAH-duhl) – taking short steps and moving slightly from side to side

INDEX

ABOUT THE AUTHOR

Eric Geron is the author of many books. He lives in Los Angeles, California, with his tiny dog. If he had to choose, he would be an emperor penguin.